Shirley da Silva Matias

Environmental Vulnerability

Shirley da Silva Matias

Environmental Vulnerability

Brief thoughts on the city of Corumbá-MS

Imprint

Any brand names and product names mentioned in this book are subject to trademark, brand or patent protection and are trademarks or registered trademarks of their respective holders. The use of brand names, product names, common names, trade names, product descriptions etc. even without a particular marking in this work is in no way to be construed to mean that such names may be regarded as unrestricted in respect of trademark and brand protection legislation and could thus be used by anyone.

Cover image: www.ingimage.com

This book is a translation from the original published under ISBN 978-613-9-64187-1.

Publisher:
Sciencia Scripts
is a trademark of
Dodo Books Indian Ocean Ltd. and OmniScriptum S.R.L publishing group

120 High Road, East Finchley, London, N2 9ED, United Kingdom
Str. Armeneasca 28/1, office 1, Chisinau MD-2012, Republic of Moldova, Europe
Printed at: see last page
ISBN: 978-620-7-73565-5

SUMMARY

The housing deficit is one of the major problems faced by the residents of Corumbá - MS, and the physical conditions of the city are a factor that limits and jeopardises urban expansion, i.e. the establishment of new areas for housing construction. In this sense, studies and research into the city's environmental structure and vulnerability are essential in order to contribute to public decision-making in the search for an appropriate coexistence between man and the environment. Thus, this work deals with the environmental vulnerability present in the city of Corumbá - MS and its main objective is to present and analyse these areas, which are scattered throughout the city and were previously classified by Pereira and Pereira (2011). The aim was therefore to deepen the study of areas with environmental vulnerability, to draw up thematic maps of the location of these areas within the city and, finally, to diagnose and assess the main impacts that the occupation of unsuitable areas can have on the population, which is generally unaware of the problem it is creating for itself. The environmental vulnerability map, combined with the analyses carried out during the field survey, is an important support tool for urban planning and serves as a subsidy for the rational use of urban land occupation, as well as for the conservation of the city's permanent preservation areas, which are currently ignored.

Keywords: Environmental vulnerability, urban planning, Corumbá, urban sprawl.

"Everyone who begins to learn about the natural sciences - sooner or later, by one means or another - comes round to the idea that the landscape is always an inheritance. In fact, it is an inheritance in every sense of the word: the inheritance of physiographic and biological processes, and the collective heritage of the peoples who have historically inherited them as the territory in which their communities operate."

Aziz Ab'Saber (2003, p. 09)

ACKNOWLEDGEMENTS

I would firstly like to thank God for the opportunity to do and complete this master's degree.

To Capes for the scholarship awarded, making it possible to carry out this research.

To the Postgraduate Programme in Geography (PPGG) at the Federal University of Grande Dourados for the opportunity to improve and enrich my knowledge.

To the professors of the Master's programme, the basis for the construction of this work.

I would like to thank my advisor Prof Dr André Geraldo Berezuk for his dedication and trust in me.

I would like to thank Professors Charlei Silva and Joelson Gonçalves Pereira for their valuable contributions to my qualification and the development of this work.

To my classmates, for the conviviality during 2012, the first year of my master's degree.

To my friends Gladslayne and Joelson for their friendship, support and welcome when I arrived in this city.

I would like to thank my friends Simone do Valle L. Peinado and Daniela Lopo for their hospitality and support during my field research in Corumbá.

I would like to thank my brother for all his support, trust and main encouragement in realising this goal.

To my sons Diego and Tiago, who, even though they are far away, are always present in my life.

SUMMARY

INTRODUCTION

In recent decades, news of catastrophes involving environmental issues has been frequent in the national and even international news. The issues related to these events have numerous causes and when they occur in urban areas they are often attributed to a lack of planning and the inappropriate occupation and use of urban land. When occupation takes place on hillsides, river slopes, valley bottoms and areas subject to flooding, the degradation process is inevitably accelerated, thus accentuating the natural vulnerability that already exists.

According to Spósito (1994), the city, as the greatest expression of the social capacity to transform natural space, is still part of nature and subject to natural dynamics and processes. However, nature is seen by a large part of the population as something that should be appropriated by man and excludes issues of cultural, social and environmental dimensions that are fundamental to maintaining a healthy life. Currently, a large part of the Brazilian population lives in cities and these, in turn, are becoming increasingly industrialised and mechanised, which is necessary in order to meet the demands of consumption and a fast-paced way of life.

In this sense, the relationship between capital and labour excludes nature, which becomes the raw material for production, making this one of the most important relationships in the modern world. However, the infrastructure needed for the well-being of the population has not kept pace with the accelerated process of urban growth experienced by most Brazilian cities.

The lack of basic sanitation, the absence of rubbish collection, open sewers, wastelands used as rubbish dumps are some of the environmental problems resulting from the lack of planning to absorb all the people migrating to the cities. The lack of jobs and the inefficient supply of housing has created a huge social disparity and the poorer population has sought to settle in places that are unsuitable for housing, such as hillsides, riverbanks and valley bottoms, with poor infrastructure and which are environmentally protected areas under national legislation. In this way, the lack of planning and infrastructure is also directly related to flooding, landslides and mudslides, which have become commonplace in recent years in many Brazilian cities.

The advance of occupation for housing purposes into unsuitable areas and disregard for geological, hydrographic, pedological and topographical characteristics, as well as environmental legislation, especially the rules relating to the Forestry Code (Law 12.651/2012), lead a portion of the population (the poorest) to greater exposure to natural events such as flooding, slope slides, landslides of large rocks, risks of heavy runoff and other environmental risks.

Thus, this problem of highly vulnerable and environmentally risky areas, when inhabited, ends up leading the population to lose their quality of life, material losses, culminating in the loss of human life itself, as a result of the intense occupation of unsuitable areas. News of these catastrophes has become increasingly frequent in different regions of Brazil, reflecting the insecurity of today's society and the inertia of public authorities. This critical panorama of urban planning is also occurring in the city of Corumbá-MS, a situation that prompted the development of this dissertation, choosing as its theme the environmental vulnerability present in this city. Therefore, the aim of this research was to diagnose and assess the main negative impacts that inappropriate occupation can have on the population and also on the city of Corumbá itself, impacts that stem from a historical lack of inertia on the part of past municipal administrations.

The relevance of this work is therefore understood to be a tool capable of subsidising new research, work and planning actions on the issue of environmental vulnerability, related to the problem of irregular occupation, especially in hillside areas and those subject to flooding. This work is organised into five chapters that discuss the subject in the following order:

The first chapter refers to the theoretical background, contextualising the proposed research topic and the theories of authors who define the issue of Environmental Vulnerability. This chapter also includes a brief approach to planning and the researcher's environmental perception.

The second chapter briefly characterises the study area and presents the physical and natural aspects as well as the socio-economic transformations that have taken place in the city and the history of the process of occupation of the peripheral areas, with the aim of portraying the existing problems in the study area, based on the Geosystemic and Ecodynamic conception of landscapes.

The methodological path of this research is presented in the third chapter and refers to the main issues for analysing environmental vulnerability within the city of Corumbá.

The fourth chapter deals with a reading and analysis of the Participatory Master Plan, relating it to the occupation and use of urban land and the government's proposals for the expansion of the city.

Finally, the fifth and last chapter presents the results of the field and theoretical research. Thematic maps are also presented to guide the reader to the location of each of the areas investigated.

CHAPTER 1

THEORETICAL BACKGROUND

One of the biggest problems facing Brazilian cities today is the sharp growth of the urban population. The city is the place where people look for change, opportunities and better living conditions, conditions that are not always achieved, as this population generally does not have enough purchasing power to live in the carefully planned areas for human settlements. However, according to Rodrigues (1994; p.11):

> "Somehow you have to live. In the countryside, in a small town, in a metropolis, housing, such as clothing and food, is one of the basic needs of individuals. Historically, the characteristics of housing change, but you always have to live, because you can't live without taking up space."

As a result, the poor end up settling in unsuitable areas with poor quality planning, building precarious housing and often resulting in informal settlements. These areas that are unsuitable for housing are generally those on the banks of rivers or on the slopes of hills and valley bottoms, which are highly susceptible to problems such as landslides, flooding, landslides, among others, especially during the rainy season, when these problems become more evident.

These problems often result in loss of life and material damage, as well as contributing to the environmental degradation of these areas, highlighting the existing conflict between development processes, the accentuation of poverty levels and the deterioration of the environment. The city, as the greatest expression of the social capacity to transform natural space, does not cease to be part of nature and subject to natural dynamics and processes (SPOSITO, 2003).

According to Carlos (2007), urban reality is facing increasingly complex problems that involve unravelling the contents of the urbanisation process today, reproducing the idea of the city as a human construction that reveals, over time, the accumulated relationship between the economy, society and nature. In this sense, the term environmental vulnerability has become a reflection of this tripod, reflecting the constant insecurity of a population exposed to danger, which is the response to the way this landscape is organised.

In order to deal with the subject of environmental vulnerability, it was first necessary to understand the relationships established between society and nature throughout the history of civilisation. Until the 1950s, when it came to nature, it was thought of in a fragmented way, separating essential elements such as geology, geomorphology, pedology, vegetation and climatology, excluding man and understanding the landscape in a descriptive way.

This conception of nature as having infinite resources, associated with the current economic model, favours the intensification of activities to exploit natural resources and disregards the impacts caused to the environment, often to irreversible proportions.

From the second half of the twentieth century onwards, debates about the inclusion of man as an element in the geosystem led to increasingly complex studies about an integrated system of natural elements (RODRIGUES, 2001 p. 78).

The geosystemic vision understands that the elements of the landscape are interconnected and its main concept is the connection between nature and society and the delimitation of territorial spaces modified or not by economic and social factors, essential elements for a geoenvironmental analysis.

When we think of vulnerability, one of its concepts can be the concept of the magnitude of a foreseeable harmful event (danger) in a given space-time, and this danger can be of human, economic or environmental origin. It is the natural sphere that joins the environmental sphere, creating a single whole, a space-time of its own, which contributes to a totality that is also unique (CAMARGO, 2012).

The origin is human when vulnerability is the result of anthropogenic actions involving deforestation, the removal of riparian forest and the occupation of hillsides. These actions favour and potentiate events considered natural, which are generally predictable phenomena and would not be at risk without human presence.

Vulnerability and risk are very close concepts, and it could even be deduced that both coexist, since vulnerability associated with danger (*Hazard)* is the very foundation of risk (VEYRET, 2007).

According to Kobiyama (2006), *hazards* theory emphasises natural aspects such as avalanches, earthquakes, cyclones and landslides, while danger characterises events based on predictable natural phenomena and risk on the probability of loss seen in a given period when a danger is imminent.

For Veyret (2007), it is important to consider social, economic and environmental factors in order to understand the existing threats and environmental impacts on the territory, since environmental risks result from the association between natural risks and risks arising from natural processes aggravated by human activity.

(2009; p.18), vulnerability is not an isolated phenomenon, nor is it a given: it is part of the very constitution of places, groups and people and needs to be understood as a process.

> "The idea of vulnerability that permeates this perspective seeks to move away from an essentially negative view. We have realised over time that vulnerability is intangible, because it is not an isolated phenomenon or a given: it is part of the very constitution of places, groups and people." (MARANDOLAJR., 2008; 2009).

Valley bottoms, hillsides, riverbanks and streams are areas with environmental characteristics that reveal the locus of fragility and therefore deserve special attention, mainly because they are the areas that are generally "left over" to be occupied by the low-income population. In this way, environmental vulnerability is related to the physical conditions of a given environment whose critical points are intensified by human actions or occupations, making it possible for social damage to occur which, depending on its magnitude, can become very acute.

Having emphasised the meaning of environmental vulnerability, it is worth introducing the classic concept of the environment. According to Christofoletti (1999), the noun environment has been used in a broad and generalised way, from the most varied fields of scientific knowledge, as well as in the different media, expressing a variety of aspects in its meaning.

However, when it comes to environmental issues, it is necessary to use this term in a more ecocentric, less anthropocentric way, because it often happens that human beings mistakenly do not see themselves as an integral part of it and interpret it in the most diverse ways, according to the social group and the problem involved, with fragmented views of nature, based on the conception that it is an infinite reserve of resources to be exploited. However, this view became unsustainable and in order to prevent the environment from collapsing, the idea of sustainability began to be disseminated in the form of the ecocentric view, in an attempt to rebuild the relationship between man and nature, initiating a change in behaviour and valuing the environment.

For Christofoletti (1999), the term environment is used to represent all the components of the geosphere-biosphere, a conceptual interpretation that is consistent with the physical environmental system (geosystem), with anthropogenic relevance prevailing. In this way, it is made up of the systems that interfere with and condition social and economic activities and can be studied using two different approaches: ecological and geographical, with the geographical approach being used to analyse space and the dimensions of human activities.

Reigota (1991) created a typology to define the term environment in three different ways:

1. The naturalist view, which sees the environment as synonymous with untouched nature, typically characterised by its natural aspects;

2. The anthropocentric view, which describes the environment as the source of natural resources for human survival and, finally;

3. The globalising vision, which integrates the environment and society.

Reigota (1991) also defines the environment as the determined or perceived place where natural and social elements are in dynamic and interacting relationships. These relationships imply processes of cultural and technological creation, as well as a historical-social process of transforming the natural environment into a built environment, separating human beings from nature.

For Santos (1994), the history of man on Earth is the history of a progressive rupture with the environment to which he belongs and, to this end, he began the process of mechanising the planet in an attempt to dominate it. Society's way of life, the increase in the world's population and changes in the mode of production have therefore resulted in a number of alterations to the natural environment. The major milestone in these changes was the Industrial Revolution in the 18th century, as it was from this advent that the process of urbanisation accelerated and, at the same time, new production techniques were perfected, bringing about significant changes in behaviour between society and nature.

Logically, this society has increasingly made use of natural resources, which Foladori (1999) defines as social metabolism, i.e. external nature being transformed into material wealth that can be consumed, enjoyed and appropriated by human society, as is the case with urban land which, under the logic of capital, becomes a commodity in the hands of the city's producers.

According to Harvey (1980), land and its improvements are indispensable commodities for the reproduction of the labour force, for the production of goods and for meeting the needs of all individuals.

This type of relationship with nature, which is one of the characteristics of today's society, in addition to technological, educational and economic advances, has in turn had "side effects", such as global warming, which stems from the excessive greenhouse effect, which in turn stems from excessive pollution.

According to Araújo (2010), the following types of environmental degradation have been identified in modern society:

- Noise pollution;

- Degradation of Natural Resources in an Environmental Protection Area;

- Degradation due to Insufficiency and/or Deficiency of Public Services;

- Degradation of Cultural Heritage Assets;

- Environmental Degradation Resulting from Mineral Exploration;

- Pollution that degrades natural resources;

- Irregular Occupation in an Environmental Protection Area.

It is important to emphasise that each type of degradation involves a set of circumstantial facts, depending on the situation identified, making the presence of public authorities extremely important, either as a controlling agent or as a mediator of human interventions.

In the face of all this negative interference by mankind with nature and the environment, and with a view to improving the quality of life for human beings and all other species, mankind's relationship with nature has been rethought. The aim is to combat this negative interference by humanity in its relationship with the environment.

For Bastos and Freitas (2010), environmental preservation policy and environmental problems must continue to be on the agenda of all segments of society concerned with quality of life and, consequently, with the environment. In order to better understand this relationship, another approach has been widely used in geography to study nature, the environment and geographical space, which is the use of the researcher's perception. Oliveira (2007) states that:

> "Since the 1970s, mainly in the United States, Canada, England, France, Australia and also in Brazil, there has been an increase in the volume of research and reflection on the problem of perception in Geography." (MACHADO; OLIVEIRA, 2007. P. 129).

Through the researcher's perceptive activity, it is possible to find out how individuals perceive the world around them. According to the author, environmental perception has demanded that society reflect more deeply on the theoretical, practical and factual equation of the environment. For Ferrara (1993), environmental perception is defined as the operation that exposes the logic of language, which organises the expressive signs of the uses and habits of a place.

Considering that environmental problems are generally associated with social problems, it is important to also seek an understanding of the facts and everyday life perceived through a holistic approach, thus including the way in which the community views and recognises the space in which it lives. Knowing the limitations of nature and the alterations caused by human actions to the natural characteristics of the surface in which they live is the first step towards understanding that not every free space can be inhabited.

This work therefore begins with a presentation of the natural characteristics of the area chosen for study.

CHAPTER 2

THE CITY OF CORUMBÁ-MS: ITS PHYSICAL AND SOCIO-ECONOMIC ASPECTS

Located in the western region of the state of Mato Grosso do Sul, on the banks of the Paraguay River, Corumbá is today the main urban centre of the Pantanal and the most important city in the region.

The city has the third largest population in the state, with 103,206 inhabitants according to the 2010 IBGE Census. The municipal territory is divided into seven districts: Sede, Albuquerque, Amolar, Forte Coimbra, Nhecolândia, Paiaguás and Porto Esperança, where various localities are located, including rural settlements, urban centres, military detachments and villages. Only two districts have urban centres: Corumbá and Albuquerque. Forte Coimbra and Porto Esperança are population centres served by some community facilities such as schools, churches, cemeteries, as well as an incipient retail trade, making them settlements (LEITE 2008; p. 26).

The occupation of the region began in the 16th century, when, with the expectation of finding gold, the area of the current municipality was explored by the Portuguese, who began arriving in 1524. The exploitation of gold and diamonds only took place in the north of the province, when during an expedition in search of the Coxiponense Indians (1719) gold was found on the banks of the Coxipó and Cuiabá rivers, thus changing the purpose of the expedition and determining the development of trade in isolation from the locality.

In 1778, the Arraial de Nossa Senhora da Conceição de Albuquerque (Corumbá's first name) was founded to stop the advances of the Spanish who arrived in search of the precious mineral. The town was elevated to a District in 1838, due to the commercial importance it became with the release of Brazilian and Paraguayan boats on the Paraguay River, connecting Corumbá to the major Platine and Brazilian urban centres and leading to the city's rapid growth.

In 1850, it was elevated to the category of municipality and, with the Resolution of 5 July 1862, it became a parish and was renamed Freguesia de Santa Cruz de Corumbá or Vila de Corumbá. In 1861, the Customs House[1] was set up in the town, attracting more and

---

[1] Customs House - Set up in 1861, the customs house was the headquarters of the Federal Revenue Service. It is currently part of the architectural complex of the General Port of Corumbá - MS.

more people because of the excellent opportunities the town offered. During the Paraguayan War (1864 - 1870), Corumbá was the scene of one of the main battles of the conflict, being occupied and destroyed by Solano Lopez's troops in 1865. The local economy was completely devastated during the war.

The small plantations that existed at the time were destroyed by the Paraguayan troops, the herds of the newly established farms in the Pantanal were decimated and navigation on the Paraguay River was interrupted. However, with the retaking of the town by Lieutenant Colonel Antônio Maria Coelho in 1870 and the expulsion of the Paraguayans, the town began to be rebuilt by the residents and only the alignment of the streets was preserved (PEREIRA, 2007).

Local development began to take off with the arrival of European immigrants and those from other South American countries, which fuelled the import and export trade in Corumba's port, putting Corumbá on the map.

as the 3° largest port in Latin America until the 1930s. The goods produced in the region were shipped to the capitals of the Platine region, and through there, the goods to be consumed throughout the province of Mato Grosso arrived. After the war, Corumbá gradually reorganised itself and resumed its growth, once again becoming the busiest port in the province (ITO, 2000). Paixão (2006, p.142) also comments on this:

> "The effervescence of commerce was such that, during this period, equipment common to the world's major centres was installed in Corumbá, such as banks, theatres, drinks factories, offices specialising in export and import services, cable telephony, thermoelectric power stations, etc."

According to Ito (2000), the beginning of the 20th century was marked by an increase in livestock farming, foreign investors and the expansion of the Casa Comercial, which often combined the functions of foreign bank representative, import-export firm and shipping company. During this period, all of the city's wholesale commercial activities were located in the Port, or lower part of the city, which due to its urban structure had the topographical division of the upper part and the lower part.

The lower part of the city was home to export and import businesses, wholesale houses and various banking establishments, the Customs House and the Revenue Office. The social elite and the retail trade settled in the upper part of the city, mainly in Calle Delamare (ITO, 2000).

The city began to expand towards the south, east and west, as the Paraguay riverbed to the north was a natural obstacle to its expansion in this direction. The river has always had a strong influence on the city's dynamics and has been a witness to the great historical changes that have taken place and, at the same time, a determining factor in its development. It was along the river that, in the second half of the 19th century, Solano Lopes' troops arrived to invade the city in the war against Paraguay, destroying and practically decimating its population. But it was this same river that allowed it to be rebuilt quickly, through the international agreement for free navigation of the Paraguay River, allowing Corumbá to develop significantly (PEREIRA, 2007).

With the arrival of the railway at the end of the 1940s, the city turned towards the south, leaving aside the river and its port, which began to experience a significant decline, thus becoming the outskirts of the city. The commercial houses, now abandoned, began to be invaded and turned into collective housing for the low-income population (ITO, 2000).

From the 1990s onwards, thanks to the practice of sport fishing, the municipality of Corumbá began to attract tourists from all over Brazil and fishing tourism began to emerge as one of the main local economic activities. Tourism began to move the whole city, expanding the hotel network, boosting trade and bringing new investments from the government. As a result of this new and very promising activity, tourist agencies, new hotels and restaurants were created. The offices of the boats and some agencies began to operate in the previously abandoned houses in the harbour.

According to Pereira (2007), the low-income families who were occupying these properties were removed to make way for the new condition of Porto Geral, as the aim was to recover the real estate value and revitalise the area. This process, known as gentrification, has taken place in other Brazilian cities such as São Paulo and Salvador, Goiás. In this case, according to Carlos (2004), it is no longer a question of production in an empty space, but of the reproduction of space in an already consolidated city, which plays into the hands of the developmentalist logic of capitalism.

2.1 Socio-economic transformations in Corumba's urban space

Since it was founded, the municipality of Corumbá has faced countless transformations in almost every area, especially in the economy and consequently in the social sphere. In the 1970s, the disorganised process of urban occupation worsened, especially with the flood cycle in the Pantanal, which brought serious problems for the region's cattle ranchers, due to the flooding events that occurred in this decade, leaving almost half of the areas that were previously occupied for the development of cattle ranching flooded (ISQUIERDO, 2010).

According to Curado (2004, p. 29) the people who lived "in the burbs[2] " of the Taquari River abandoned their land and migrated to Corumbá, occupying the outskirts:

> "Other reports show that the PA 72 settlement, in the municipality of Ladário (MS), and an occupation formed by "homeless" families on the outskirts of Corumbá are home to former landowners from the Bracinho colony, and the Miquelina and Rio Negro communities, who have abandoned or sold their land to third parties."

During these periods of flooding, Corumbá would become marooned, relying only on navigation and the railway to get to and from the city, which once again fell into economic decline.

As a result of the crisis in the cattle industry in the 1970s, which bankrupted large cattle ranchers, the city was forced to look for new ways to stimulate the economy in the region, which until then had depended almost exclusively on this activity. With the motto "nature within everyone's reach", the region began to capitalise on its main commodity: the exuberant and unusual nature of the Pantanal (FATMÁTO, 2005):

> "Contemplative, cultural and ecological tourism is being encouraged, replacing fishing tourism, and farms are starting to make road trips, the so-called "photographic safaris", which are in great demand. In 2001, with the inauguration of the bridge over the Paraguay River, which represents a major advance in the infrastructure of the road transport system, Corumbá's integration with the rest of the country is consolidated. "

In search of new alternatives, local traders took part in the Pact for Citizenship and, among other things, demanded the establishment of a free trade area due to its location on the border with Bolivia and the reactivation of the Bauru - Corumbá - Campo Grande trains.

---

[2] Breaching and flooding of river banks. The main consequences of this phenomenon are permanent flooding, diversion of the riverbed, loss of productive areas, changes and losses in biodiversity, changes in the flood pulse and changes to navigation channels.

Despite the economic instability it was experiencing, the city continued to grow, albeit irregularly. The results were suffered by the less favoured population who, as a result, sought housing in the outlying areas, a process we will discuss below.

2.2 The process of occupying the periphery

According to Rodrigues (2001, p.29), one of the ways in which the poor solve their housing problems is by settling in the outskirts, usually in an irregular and clandestine manner. Thus, at the same time as better standard housing was being built in Corumbá's city centre and in the area corresponding to Porto Geral, most of them with European characteristics, a process of occupation of the outskirts was beginning in the areas at the eastern and western ends of Porto Geral.

According to Souza (2003), the *Correio do Estado* newspaper, on 16 June 1909, pointed out that there was a shortage of houses to cater for the population and that families were forced to share a single room with another, damaging their health and hygiene. The eastern end of Delamare Street, which is now the Borrowiski neighbourhood, was occupied by Corumbá's black population, who had very low purchasing power. At the time (1909), they formed the Sarobá neighbourhood, which was known as a place for misdemeanours and cheap labour. This was a population that was visibly marginalised by the white-skinned population with higher purchasing power, a population that considered the place a disgrace to the city. These areas were considered to have no commercial value, also due to their extremely rugged topographical conditions.

Thus began the historical process of illegal occupation of the hillside areas by poor families, made up of fishermen, former slaves and Paraguayans who were mostly survivors of the Paraguayan War. According to Souza (2003), in addition to the Sarobá neighbourhood, at the western end of Delamare Street, between Cáceres and Cervejaria slopes, soon after the end of the Paraguayan War (1865-1870), the Acampamento de Cima neighbourhood (now the Cervejaria neighbourhood) was formed, originating from the camp of the soldiers who took part in the war and stayed there. At the time, as these were temporary dwellings, they did not follow the original layout of the city plan, a fact that was already contested by the City Council, which was discussing expropriating the area and bringing it back into line with the original city plan.

The layout and organisation of the streets, in the shape of a chessboard, only covered the central area, and this planning did not take into account the physical characteristics of the city, favouring only flat areas. Also according to Souza (2003), since the 1900s there had been a lack of housing units in the urban centre of Corumbá for the poorer population. This fact was visible in the very landscape of the city, which therefore revealed itself in a segregationist and exclusionary urban design.

Occupation of the upper part of the city only began in the 1940s with the expansion of the urban area to the south, creating new neighbourhoods in that region. Due to the distribution of morrarias, these neighbourhoods were created in the existing valleys which, according to Pereira (2007), have a topographic pattern of downward sloping planes, from the base of the hills towards the talvegue lines, favouring occupation.

According to Pereira (2007), the first valley is occupied by the Cristo Redentor neighbourhood and the Vitória Régia and Cravo Vermelho I, II and III housing estates, surrounded by the Cruzeiro hills to the north, Pico de Corumbá to the west and Bocaina to the south and southeast.

The occupation of the city's peripheral areas was not only the choice of the residents who sought out the less valued areas to settle in, but was also induced by the government itself, which built housing estates and popular housing developments in the morraria areas. This situation, which began in the 1970s, continues to the present day, with the omission of municipal managers who, out of political interests and complete disregard for the needy population, recently approved the construction of 1,200 houses as part of a state government project in areas subject to flooding and already earmarked in the masterplan for the construction of industries.

The rural exodus and the settlement of new areas in Bolivia with the implementation of commercial and industrial activities in the 1990s, as well as the boom in tourism, were also factors that contributed to the increase in the urban population, bringing with it a major problem: occupations in new irregular and/or unsuitable areas for housing, this time advancing to the slopes of the hills and hills located to the south of the city.

2.3 Physical aspects of the urban area of Corumbá-MS

The physical characteristics of a study area, as well as its socio-economic characteristics, are extremely important in order to understand and analyse the vulnerability of the place. In this way, we also consider the anthropogenic activities carried out and, above all, how the place is reacting to these activities and what the results of these interferences are for both the environment and the population. Following this reasoning, better planning is possible with more precise decision-making.

Table 1 below summarises the lithostratigraphic units associated with the geological and geomorphological features found on the Corumbá urban site. The table follows:

GEOLOGY	Two lithostratigraphic units in the urban area: 1-0 Corumbá Group represented by the Bocaina Formation with predominantly dolomitic limestones and the Tamengo Formation - top units, in which calcitic limestones predominate and fossiliferous content is abundant. 2-Pantanal Formation - Formed by continental clay deposits (fluvial, lacustrine and/or fluviolacustrine), it is subdivided into 3 units: Qp1 - Semi-consolidated sandy conglomeratic sediments; Qp2 - Semi-consolidated sandy-clay sediments Qp3 - Semi-consolidated sandy-clay sediments
GEOMORPHOLOGY	Relative Depressions - The predominance of flat terrain associated with the effects of urbanisation has serious limitations for infiltration and surface drainage. Slopes - Steep terrain that stretches along the entire riverside of the urban area. It has a slope of more than 50% and a height of between 20 and 30 metres. Valley bottoms - Stretches under the influence of slopes with a degree of drainage deepening. Inselbergs - Collinear elevations of the Bocaina Formation, with altitudes ranging from 150 to 453 metres and a slope of more than 30%. Accumulation Modelling - Flattened terrain occupied by the Pantanal formation, it extends throughout the region bordering the urban area. Pediplano - Flattened surface with slopes of less than 5%

Table 1 - Summary of the geological and gemmorphological feature units present in the Corumbá-MS urban site. **Source:** Pereira and Pereira (2010) **Organisation:** Matias, Shirley S. (2014).

2.3.1 Geological and geomorphological aspects of the urban area

The city's geological framework is made up of the Corumbá Group, through the Tamengo and Bocaina formations. This group is predominantly characterised by denudational relief on the plain and moderately dissected hills. The dedudational/structural reliefs are located mainly in the southern sector of the city and are made up predominantly of morrarias (ISQUIERDO, 2010).

The name Tamengo formation derives from the Tamengo Canal, which connects the Paraguay River to the Cáceres lagoon (Bolivia), the area where

this formation is located. This unit was defined by Almeida (1945) and covers an area of approximately 90 km^2 and is restricted to the city of Corumbá.

According to the author, this formation occurs on the right bank of the Paraguay River, in the stretch that encompasses the urban area of Corumbá and Ladário. Its fossiliferous content is the richest among the formations of the Corumbá Group and has been the subject of work by several authors who indicate a possible Neoproterozoic age for this formation.

The geological base on which the city of Corumbá is built behaves geomorphologically like a slope descending in a south-north direction and, due to this peculiar formation, combined with the geological and geomorphological characteristics, it has a very rugged relief with different topographical compartments, which imposes serious limitations on its expansion.

> This relief was formed over the carbonate rocks of the Bocaina and Tamengo formations of the Corumbá group, dominantly marine sedimentation, which based on lithological and stratigraphic comparisons is accepted as being Proterozoic in age. The thickness of these deposits can exceed 400 metres in some areas (ISQUIERDO 2010 p.5).

The city has two morphologically distinct environments, which place it on two levels: the first, known as the lower part of the city, is marginal to the Paraguay River and developed on an old flood plain directly subject to the hydrological dynamics of the river channel, represented by its annual cycles of floods and ebbs (PEREIRA, 2007).

The Bocaina Formation was named after a river gorge of the same name, located 4 kilometres southeast of the city of Corumbá. It is a geological formation from the Pre-Cambrian period and is the predominant surface formation in the Corumbá Group area. It outcrops on dolomitic hills and, subordinately, on calcitic limestones and marbles (ALMEIDA, 1945).

The natural geological erosion of these sediments has therefore exposed limestone rocks from the Bocaina/Araras Formation in a large part of the region, consisting of limestone, dolomitic limestone, marl limestone and marble (BRASIL, 1979). Figure 01:

Fig. 01- Exposed limestone rocks of the Bocaina/Araras Formation. **Photo:** Shirley Matias, 2013.

According to Cunha & Guerra (2003), the pediplanar areas constitute an extensive flattened surface with slopes of less than 5% between the Paraguay River and the inselbergs of the Bocaina Formation. The geological characteristic of this formation is limestone rock with the presence of flattened hills. Figure 02:

Fig. 02 - Pedestrianised areas with a slope of less than 5%. **Photo:** Shirley Matias, 2013.

The northern sector of the city of Corumbá has a very steep slope, with large rock formations that can be seen outcropping at various points in the city, as well as the presence of wetlands and pedologically unstable areas along the Paraguay River.

Fig. 03-Rock formation on the riverside slope in the Cervejaria neighbourhood. Photo: Shirley Matias, 2013.

According to Pereira and Pereira (2010), six dominant topographic compartments were identified in the urban area of Corumbá: valley bottoms; relative depressions; pediplanar areas; inselbergs; accumulation modelling and the riverside slope.

25

2.3.2 Pedology

According to Embrapa Pantanal's Technical Circular No. 18, the soil in the city of Corumbá has more than one edaphic class due to the different geological and geomorphological characteristics, presenting a shallow soil with a horizon varying in colour from black to grey, which is called Rendizina.

Non-crystalline limestone, sedimentary or metamorphic rocks are porous and permeable, which means that water remains in the profile throughout the wet season. Figure 04:

Fig.04 - Porous and permeable limestone rocks on the hillsides.
Photo: Shirley Matias, 2013.

The colluvial sediments of the ferruginous sedimentary surfaces, which are deposited on the limestones on the slopes, form extremely rocky, stony and gravelly soils (CUNHA, 1985). Figure 05:

Fig.05 - Rocky and gravelly soils on the slopes.
Photo: Shirley Matias, 2013.

According to Monteiro (1997), the area is also home to shallow, poorly developed lithic soils with a clayey texture and very low porosity, which severely limit their infiltration potential, leading to greater rainwater run-off. These characteristics can be seen in figure 06 below:

Fig. 06 - shallow, poorly developed lithic soils with a clayey texture and very low porosity.
Photo: Shirley Matias, 2013.

The floodplain is also home to Grumosols[3] , which can occur simultaneously with salinised or alkalinised areas. They are made up of expansive clays and absorb organic matter on their surface.

2.3.3 Natural drainage

The city's natural drainage is characterised by the predominance of the main supply source, the Paraguay River and the Tamengo Canal, which border the northern perimeter of the entire urban area. In addition to the Paraguay River, the natural urban drainage network is made up of micro-basins and intermittent and perennial streams that flow into the Paraguay River.

There are a total of 11 urban basins with varying shapes and lengths that make up the urban area's natural drainage network, one of which extends beyond the urban perimeter. These basins and streams can be seen on the thematic map where they have been spatialised (map 01).

---

[3]Grumossolo - Typical of flat topography; clayey; good fertility; medium thickness.

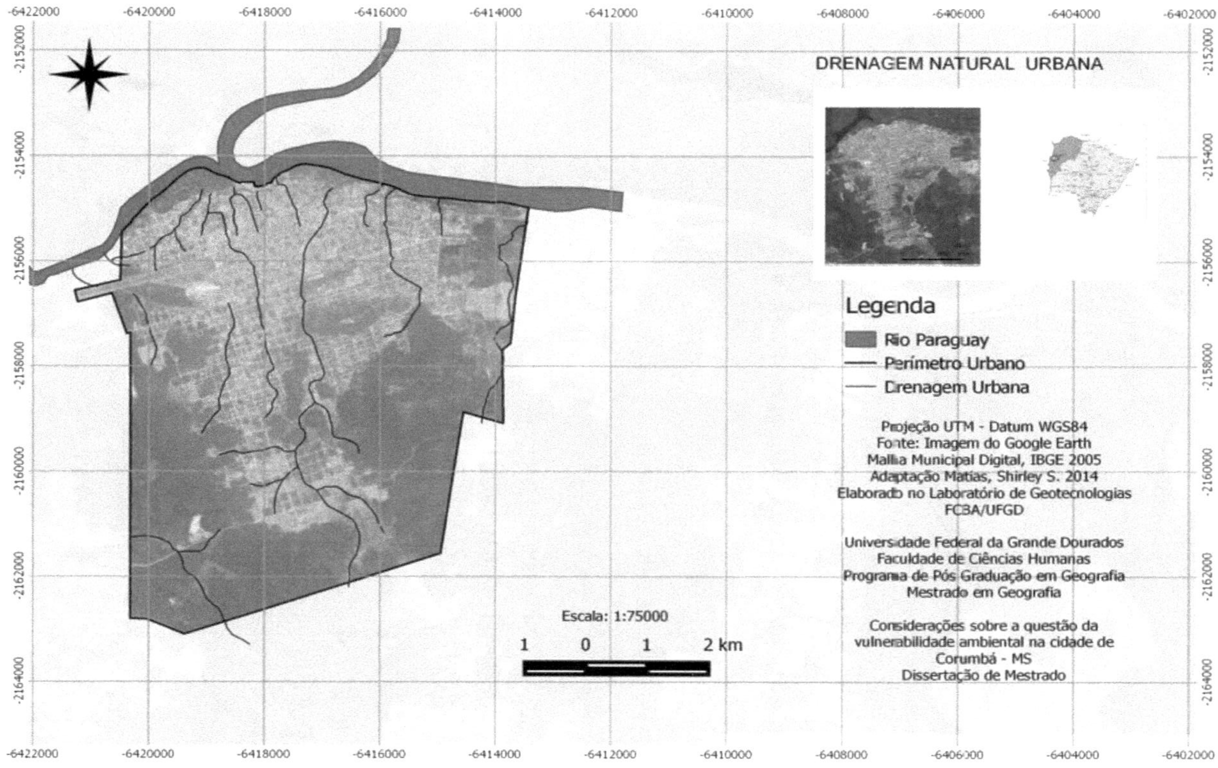

Map 01 - Natural Drainage of the City of Corumbá - MS. Adaptation: Mataa; S. Shirley. 2014.

The lack of an urban drainage chart makes it difficult to plan and comply with the municipality's land use law. As a result of the city's urbanisation process, some basins have lost their main morphometric shape and some streams have been diverted and/or landfilled to make better use of the area, resulting in serious problems for the local population. According to Matias (2013), the natural characteristics of the urban settlement, such as the geological site, geomorphology and soil type, are not being taken into account and respected in the occupation and expansion of urban land.

It is worth remembering, however, that Federal Law No. 6.766 of 19 December 1979 provides for Land Use Parcelling and, among other provisions, forbids land parceling in "swampy land and land subject to flooding, before measures have been taken to ensure the flow of water". Here is the text of the law:

> II - on land that has been landfilled with material harmful to public health, without being previously cleaned up;

> III - on land with a slope equal to or greater than 30 per cent (thirty per cent), unless specific requirements of the competent authorities are met;

> IV - on land where geological conditions do not recommend building;

> V - in ecological preservation areas or in areas where pollution prevents acceptable sanitary conditions, until corrected.

The lack of supervision by the city council favours the occupation of these areas in disagreement with the Master Plan and the Municipal Law on Land Use and Occupation.

Urban land. Due to its natural characteristics, the urban area of Corumbá has some limitations to its expansion, since the presence of large hills such as Bocaina, Bandeira, Cruzeiro and Pico de Corumbá, all located on the urban site, are geographical accidents specific to the city and are treated in the Master Plan as Special Areas of Environmental Interest.

In this way, the city's Master Plan uses the term AEIS (Areas of Special Social Interest) to replace the term ZEIS (Zone of Special Social Interest), normally used in master plans, with only the already urbanised area being considered for the purposes of land occupation.

According to the Evaluation Report of Corumbá's Participatory Master Plan, approved in 2006, due to its location in the "heart" of the Pantanal, the environmental issue should have been the central theme of this document. However, the issue was not covered, nor were the serious environmental problems that exist, especially in relation to the high basic sanitation deficit, a fact that in recent years has been minimised through projects linked to the Growth Acceleration Plan (PAC).

Currently, the city government is trying to solve the problems of homelessness through public policies, with the housing programmes of the federal and state governments, a topic that will be addressed in the next chapter.

CHAPTER 3

RESEARCH METHODOLOGY

The main aim of this work is to discuss issues relating to environmental vulnerability in the urban area of Corumbá-MS, as well as to produce cartographic products of the most critical areas in order to collaborate with urban planning in these areas. The nature of this objective allowed for a qualitative methodological approach to the research, as this defends the idea that understanding and interpreting social and human phenomena are directly related to describing the object of study.

According to Godoy (1995, p.62), qualitative research is characterised by having the natural environment as the direct source of data and the researcher as the fundamental instrument, as well as its descriptive nature and the possibility of using the perspective of landscape analysis and phenomenology. In this way, the qualitative methodology allows us to come to an understanding and interpretation of the issue of environmental vulnerability in the city of Corumbá-MS

This research was therefore based on the data presented by Pereira and Pereira (2010), seeking to extend and complement the information obtained by these authors *in loco.*

The study of the concept of vulnerability permeates various fields of study and is currently also present in the field of Geography. In order to support this work, a lot of reading was done on the subject with the aim of providing a theoretical basis for the research. The concept of natural vulnerability, related to the physical characteristics of the place and anthropogenic influences, based on the principle of Ecodynamics by Tricart (1977), is of great importance for identifying and characterising areas exposed to environmental vulnerability, for a holistic approach to the situation.

After the bibliographical survey stage, the research required the necessary data collection to understand the characteristics of Corumbá's urban area. In this way, a field survey was carried out to ascertain the conditions of vulnerability in the city, these areas having already been classified through the work of Pereira and Pereira (2010). This field survey of the study areas was fundamental for understanding the *status quo of the* issue of Corumbá's areas of high environmental vulnerability, as well as for later analysing the urban planning policies being implemented in Corumbá.

After the bibliographical survey stage and, subsequently, the field survey and photographic recording of the areas, the third stage of the dissertation consisted of drawing up thematic maps of vulnerability, declivity and micro-basins in Corumbá. The Quantum Gis 1.8 and 2.0 programmes were used to create these maps, based on the data provided by Pereira and Pereira (2010). This programme was chosen because it is free. The maps were produced at the Geoprocessing Laboratory of the Faculty of Biological and Environmental Sciences - FCBA - of the Federal University of Grande Dourados. The data for the slope map was obtained from Embrapa's official website.

3.1 - Interviews and Theoretical Assumptions of the Participatory Masterplan

The fourth stage of this dissertation involved interviews with Corumbá government officials directly involved in the city's urban planning activities. After surveying the field and reading up on the city's public policies, two city managers were interviewed in order to obtain information about the situation in the study areas. For this, it was necessary to first contact the city's Planning Manager, who clarified the situations regarding the planning and organisation of the city, followed by a second interview with the housing coordinator, who also reported on the proposals for the city of Corumbá. The interviews were conducted in a semi-structured way and took place at the Corumbá City Hall, in the Infrastructure Secretariat - SEINFRA.

At this stage, priority was given to managers rather than the population, as the aim of this work was to find out what guidelines the city council intends to follow in order to solve or minimise the problems of families living in areas of environmental vulnerability. Likewise, greater emphasis was placed on areas with medium and high vulnerability, as these were considered to be the most aggravating situations.

The analysis of the areas studied was based on the technical readings carried out during the research and sought to relate them to the reality encountered and the policies implemented in the city. The aim was to analyse the actions of the public authorities in relation to the areas presented and to present the impacts that are being caused in these areas and the reflection of these actions on the residents of these areas themselves.

CHAPTER 4

CORUMBÁ AND ITS PUBLIC POLICY ON LAND USE AND OCCUPATION

Public policies are understood to be a set of programmes, actions and activities developed by the state directly or indirectly, with the participation of public or private entities, which aim to guarantee a certain right of citizenship in a diffuse way or for a certain social, ethnic or economic segment. These public policies are created to respond to the demands of different sectors of society, especially those considered to be vulnerable or marginalised. They can be formulated mainly on the initiative of the executive or legislative powers, separately or jointly, based on the demands and proposals of society in its various segments.

Through the Municipal Organic Law and other legislative instruments, the 1988 Federal Constitution gives municipalities autonomy to expand their competence in important areas for planning, such as urban policy. To make this possible and define municipal competence, some guiding instruments have been created, such as the Master Plan (Law 10.257/2001 - Statute of Cities), which is compulsory for cities with more than 20,000 inhabitants, the Urban Land Use and Occupation Law, which can be found in the Municipality's Code of Postures (Law 004/1991) and the granting of land use and Urban Usucaption.

The Municipal Master Plan directs the function of urban areas through zoning (expansion zones, urbanisable zones and special interest zones), all with the aim of fulfilling basic functions such as: living, working, recreation and getting around, and spatially distributing socio-economic activities within the urban area. Unfortunately, what is established in the plan is not always complied with and, more often than not, public policies end up favouring the economic aspect more than the social one. In addition to the Participatory Master Plan established by Complementary Law No. 098/2006, the municipality of Corumbá has some extremely important regulatory instruments to guide its urban growth, which have been organised in the table below:

Table 01 - Regulatory Instruments of the Municipality of Corumbá-MS

Law	Year of Creation	Objective
Municipal Law No. 648	1972	Institutes the Municipal Works Code and establishes Urban

		Development.
Organic Law	1990	It meets the interests and peculiarities of the municipality.
Law No. 004	1991	Establishes the Code of Postures of Corumbá
Law no. 1421	07/08/1995	Creates the Municipal Environment Council - CM MA and the Municipal Environment Fund - FMMA;
Complementary Law No. 023	27/05/1997	Creates the Municipal Department of Environment and Tourism.
Complementary Law No. 098	2006	Institutes the Participatory Masterplan
Municipal Law no. 2.097	29/072009	Grants tax incentives to builders who carry out projects linked to the Minha Casa Minha Vida Programme
Municipal Law No. 2.275	14/11/2012	Adds the sole paragraph to art. 3° of Law no. 2.097/2009

Source: Corumbá City Hall. **Organisation**: MATIAS; S. Shirley (2014).

According to Corrêa (1989; p. 7), the urban space of a capitalist city is made up of different land uses juxtaposed to each other. The articulation between these different uses, which are basically linked to the economy, leisure and commercial relations, is manifested by public authorities and their ideologies.

> "It is a social product, the result of actions accumulated over time and engendered by agents who produce and consume space. The action of these agents is complex, deriving from the dynamics of capital accumulation, the changing needs of reproducing relations of production and the class conflicts that emerge from it." (CORRÊA. 1989; p. 11).

In Corumbá, it is possible to recognise, in the city's very landscape, all of its history and the transformations that have taken place as a result of the oscillations of its economic phases. The large mansions, which are currently listed as heritage buildings, reflect the period of exuberance and economic power of the time, while the irregular and clandestine occupations show a period of crisis and impoverishment of the population. The city of Corumbá suffered for many years under the influence and particular interests of its rulers, who did nothing to benefit the city and its population.

Inspired by the populist style of politics promoted by Getúlio Vargas, some political leaders abused personal charisma and ignored the wishes of the population in favour of the wealthier classes. There are reports from residents that one of these leaders, for example, went so far as to subdivide land on the slopes of the hills in order to give it to the poor in exchange for votes in the elections, in total disregard of the urban land subdivision law and Federal Law 6.766/79, which does not allow subdivision on land with a slope equal to or greater than 30%, unless the specific requirements of the competent authorities are met (and logically these requirements were ignored), considering constructions on slopes greater than 30% to be at risk.

In the same way, the land on the riverside slopes was allotted, making it difficult today to remove these residents who have become heirs and owners in the area.

During the field survey, it was possible to have quick conversations with the residents of the riverside region, who clarified the reason for their resistance to leaving the area, as they claim to have the land and housing documentation regulated by the town hall itself, documents registered with a notary's office. What's more, they refuse to leave the site even though they are aware of the risky situation to which they are exposed, because they invested in buying the land, building the houses or inherited them from their parents or grandparents. They also said that the houses to which they will be relocated will not be free, there is a small amount to be paid and not everyone wants or can afford it. For the residents, this is an injustice, since in a way they are being forced to buy a property that doesn't interest them. Another complicating factor in the situation is the location of the new homes and their jobs, as most of them are well-established in their professions as fishermen, bricklayers, electricians, etc. and have a clientele in the city centre and in neighbouring districts such as Generoso and Arthur Marinho.

The property sector also contributed to the occupation of unsuitable areas, as land in regulated areas with better building conditions was overvalued.

Corrêa (1989; p.23) also points out that the spatial action of property developers is uneven, creating and reinforcing the residential segregation that characterises the capitalist city. This situation of segregation is reflected in the creation of housing estates, especially those launched in recent years by the federal government through the "Minha Casa, Minha Vida" programme, with the aim of relocating some of the families living in risk areas, transferring residents who previously lived close to the centre to an isolated area (figure 07):

Figure 07 - Residential complex built with PAC funds.
Photo: Shirley Matias, 2013

This expresses the power of the state to act as a shaping agent allied to the interests of the dominant classes and social elites in the reorganisation of urban space:

> "If urban space is the product of unequal relations, these transformations, directed by public authorities, also take into account economic interests, almost always conditioned by property speculation. Human needs, in this context, are left in the background, behind the concern for the profit of commercial activities." (MATIAS;Shirley,S. 2013,p.9)

The expansion of the city into these areas far from the centre and urban facilities ends up generating problems that were non-existent before, such as distance from work and difficulty in accessing public services, thus creating a new periphery and triggering a process of social vulnerability in Corumbá. It is worth noting that Municipal Law 2.275 of 14 November 2012 adds a single paragraph to article 3° of Law 2.097/2009, authorising the executive branch to grant tax incentives to builders who carry out projects linked to the "Minha Casa, Minha Vida" programme, confirming the above statement.

4.1 About the Participatory Masterplan

The Federal Constitution of 1988 and the City Statute Law of 2001 sought to strengthen municipalities and give them more autonomy as a result of the countless political, economic and social transformations the country was facing in the 1980s and 1990s. The public policies that existed during this period were not enough to meet the needs that the process of city formation and development was demanding, because the disorganised growth in many Brazilian cities was so rapid that the so-called "informal cities" were created, produced by the poorest population, who tried to solve their housing problems by buying plots in areas on the outskirts, usually in clandestine subdivisions (RODRIGUES; 1994, p.29).

The numerous demands resulted in the creation of Law 10.257/01 - the City Statute - to regulate articles 182 and 183 of the 1988 Federal Constitution. As a result, cities with a population of more than 20,000 inhabitants are obliged to draw up and implement a Masterplan. However, most of the time, the plan is only drawn up to comply with the legislation and remains shelved for years, according to the interests of its managers.

The implementation of the Master Plan was a way of making municipalities responsible for planning, while also obliging them to fulfil the social function of the city, as provided for in articles 182 and 183 of the 1988 Federal Constitution, which form the chapter on Urban Policy. It must contain all the necessary adjustments, as well as urban planning measures to meet the city's needs.

According to Oliveira (2012; p. 63), territorial management presupposes an interaction of spatial actions, with regard to the use and occupation of space, considering the natural, social and economic attributes that involve the whole of society, essential requirements of the masterplan, which must also address

the expansive processes and their economic and social repercussions on the city, avoiding inequalities, exclusions and social segregation. Thus, with the implementation of this management tool, each city has its own guidelines, taking into account its natural characteristics and reflecting the local identity. The document must be drawn up in accordance with these specificities and be geared towards the rational development of urban space. It must be drawn up with the contribution of the local population, as this would guarantee that it fulfils popular demands and also contributes to improving the population's quality of life.

4.2 Corumbá-MS Participatory Masterplan - From Theory to Practice

Following the reasoning behind the aforementioned premises, Corumbá's Participatory Master Plan was drawn up in 2006, during the administration of former mayor Ruiter Cunha de Oliveira (2004-2012), of the Workers' Party. The plan is still in force under the current administration of Mayor Paulo Duarte, also from the PT.

After eight years since this Master Plan was enacted, it's interesting to take a look at this instrument, noting the critical and/or neglected points in its fulfilment. It is also important to emphasise that, during the process of drawing it up, various workshops and public meetings were held with the aim of bringing the local community on board to collaborate in the drafting of the plan. During the drafting period, interviews were held with the presidents of neighbourhoods, residents and representatives of class organisations. The workshops were open to the community and the meetings were decentralised, so that they could cover all the neighbourhoods and the people of Corumba could participate more actively, presenting proposals and discussing local needs.

Meetings were also held in the settlements and districts. Thus, the drafting of the Corumbá-MS Participatory Master Plan fulfilled all the recommendations set out in the City Statute, in its technical form, in order to fulfil its main objective, which is to build a city with urban quality for all, seeking to avoid and eliminate the formation of slums, irregular or informal housing.

Although, in theory, the administrative instrument contains all the necessary guidelines for balanced management, it can be seen that some of its proposals have not been put into practice and some of its problems have taken a back seat. The city still faces serious problems when it comes to its urban policy.

The existing disagreement can be seen in Article 6° of Chapter II, which refers to the general objectives. Paragraph II refers to spaces for housing production aimed at lower-income social segments, prioritising central areas, as well as the urbanisation and land regulation of areas occupied by low-income populations, with a view to the social inclusion of their inhabitants.

The opposite is true: the low-income population is being driven to the outskirts, completely cut off from the urban facilities needed to carry out daily activities, since these new housing estates still lack public services, social facilities, leisure spaces and, above all, are very far from employment centres (a fact that justifies and reinforces the return of residents to their old homes and the abandonment of the current one).

Among the guidelines set out in Corumbá-MS's Participatory Master Plan is the Land Use and Occupation Planning Law, which should have been drawn up to regulate occupations within the urban space, but so far has not. The municipality has also yet to implement its Municipal Environmental Policy, making it difficult to plan, supervise and manage the areas that are defined as AEIS - Special Areas of Social and Environmental Interest.

According to Title IV, "Final and Transitional Provisions", in Article 62 of the Master Plan, it should be reviewed within five (5) years of its enactment, and according to Article 63, the executive branch should submit the following projects to the city council within one year of the Master Plan's approval:

I - Land Use and Occupation Planning Law;

II - Revision of the Municipality's Code of Works and Postures;

III - Municipal Environmental Sanitation Policy;

IV - Municipal Urban Mobility Policy;

V - Municipal Housing Policy;

VI - Delimitation of Indigenous Areas;

VII - Municipal Environmental Policy;

VIII - Municipal Cultural Heritage Policy;

IX - Delimitation of the Albuquerque District Headquarters.

Article 65 of the plan also provided for the following instruments to be drawn up within a year:

I - Geotechnical map of the municipality;

II - Urban Afforestation Programme;

III - Planialtimetric Mapping of the Municipality's Urban Area;

IV - Average Ordinary Flood Line of the Paraguay River, in the Urban Perimeter;

V - Water, Forestry and Mineral Resources Management Plan.

These instruments would be fundamental for the proper development of the city and the well-being of the population, but some of them were simply ignored and others didn't even make it off the drawing board. What can be seen is that the provisions related to the economic part of the plan have all been implemented to some extent, such as the progressive IPTU or the Municipal Cultural Heritage Policy, which favours tourism in the region and encourages the economy to the detriment of social and environmental issues. In recent years, a lot of investment has been made in the city, especially in the construction of housing estates and tourism-related projects, such as the Miguel Gómez Convention Centre, built on the quayside of the general port on the River Paraguay with the aim of boosting business tourism in the city. There was also the paving of some neighbourhoods on the outskirts, through the "Se essa Rua Fosse Minha" project, which used the labour of local residents to carry out the project.

Other aspects remain deficient, such as public transport, which is currently in precarious conditions and has been declared a state of public calamity by the current administration. According to decree 1.137/2014 published in DIOCORUMBÁ, the mayor points to the lack of maintenance of vehicles, the inefficient number of buses on the routes and the risks to the physical integrity of users. Rubbish collection is also a problem in the city. Recently, the Federal Public Prosecutor's Office (MPF)[4] in Mato Grosso do Sul called on companies and public bodies in the municipality of Corumbá to guarantee the right of approximately 100 residents of the Campos Correia Family Quilombola Community to have access to basic assistance services, because the community does not receive a collection service and rubbish accumulates, favouring the proliferation of vectors. However, it should be noted that the middle-class neighbourhoods and the city centre receive a daily cleaning service, while on the outskirts the service is still precarious.

In this way, we realise that the Participatory Master Plan of Corumbá-MS, despite being an instrument of great importance for the management of the city, is not corresponding to the wishes of the population. It is necessary for all its guidelines to be complied with, especially the social and environmental issues that are simply being ignored by the government.

4.3 Analysing the interviews

The representative[5] said that the plots had been acquired irregularly and that there were long-term projects to be implemented by the city council for both the riverside areas and the hillsides. He presented the now well-known Corumbá Waterfront Project, which is to be implemented in the riverside region, with the aim of promoting sustainable development in that area. According to the interviewee, the plan is to remove the families settled in the most critical areas of vulnerability and keep some, with the construction of some three-storey, regulated housing units so that the families remain in their original location.

The housing manager also said that the construction will be financed by FONPLATA (Brazilian Fund for the River Plate Basin) and, according to the project, the housing units will be built away from the slopes and away from the river's flood zone. He also made a point of saying that the Orla Corumbá project is still in its infancy and that some

---

studies will have to be carried out before it can be put into practice.

With regard to the implementation of the works under the Growth Acceleration Programme, in terms of infrastructure it will provide a major improvement in the drainage of the entire urban area and a linear park will be built on the slopes of the hills, which are currently occupied irregularly.

There are plans to build 1,600 housing units, including houses and blocks of flats of up to three storeys each, to make up for the city's housing shortage and also to accommodate families who will be removed from risky areas and occupations on hillsides. The area for the construction of these housing units has already been defined and is in the process of being cleared and levelled.

Regarding urban expansion, it was reported that the only forecast for the next 20 or 30 years would be to expand the city towards the settlements, which are already quite close to the urban area. Given the city's characteristics and the limited area available for expansion, the most likely alternative is to extend the urban area towards the Taquaral settlement.

The municipality also intends to implement economic and social incentive policies, bringing a new economic centre to these new housing estates, with the construction of markets, petrol stations, bank branches, schools, pharmacies, crèches, etc. To this end, Complementary Law No. 160 of 17 September 2013 was created, which provides for the policy of incentives for economic and social development in the municipality of Corumbá and creates the Corumbá in Development Programme (CODES).

New professionals in the fields of psychology, social work and environmental engineering are being appointed to take part in a committee that will provide emotional support so that residents don't feel excluded in their new homes, given that many of them are returning to their old homes, even though they know the risks they are exposed to.

Regarding the urban voids that exist in the city centre, the question was asked whether they could be occupied for the construction of social housing and it was reported that there are a large number of urban voids, but that the city council intends to apply the progressive IPTU system, as provided for in the masterplan, and after five years the property will be expropriated.

The conversation with the two city hall managers was nothing new, as they both recounted familiar situations that have gone on for several years without any solution. What can be seen is the power of the state prevailing over the interests of the population, and with a powerful weapon - the supply of housing.

The urban growth actions carried out by the public authorities feed the process of peripheralisation and exclusion, including the different construction standards and housing locations, building housing estates completely outside the commercial axis of the city and others closer to it and in more populated areas, but with very poorly structured and comfortable buildings, in valley bottom areas, without asphalt or any kind of interior finish. In this way, the electoral interest and lack of sensitivity towards the residents who, for the most part, have no alternative but to accept what is offered to them is obvious.

The housing estate, built in an area designated by the Master Plan as the city's industrial sector, was for some time a source of conflict between the municipal and state authorities. Despite all the irregularities, the complex is still inhabited and the residents are suffering the consequences of electioneering.

So, once again, the population is both the victim and the object of political and economic interests and those in power.

The photos below show the disastrous situation found during the field survey (Figures 08 and 09).

Fig. 08 - Partial view of the housing complex built in a valley bottom area.

Author: Shirley Matias,2014

Figure 08 shows the precarious infrastructure of the plots, the lack of tarmac and the saturation of the soil. Because of the poor housing conditions, some families have abandoned their homes and moved back to where they used to live.

Figure 09 below shows the resident with the furniture and objects in her house upended due to the imminence of flooding, a fact that occurs every time it rains in the city, as reported by the resident.

Fig. 09 - Interior of a house in the state housing estate.

Author: Shirley Matias, 2014.

The new housing estates located practically outside the city, already on its ring road, is yet another confirmation that commercial, political and economic interests are above social ones. Working with professionals to get the residents to accept their new status as outcasts is nothing more than forcing them to accept what is being imposed on them, making them believe that this is the best solution to their problems.

For Lefebvre (2004, p.116), the right to the city cannot be conceived as a simple right to visit or return to cities, and can only be formulated as a right to renewed and transformed human life, as opposed to the new housing being built.

Regarding the large number of urban voids reported during the interview and their underutilisation, serving only to proliferate vectors, open-air rubbish dumps and, above all, encouraging real estate speculation, these are in the central and consolidated area of the city, occupying the best spaces and with complete infrastructure. However, it is not in the administration's interest to use them for decent housing projects for low-income families. In short, the public authorities are showing less and less interest in maintaining, or at least trying to maintain, a good understanding of a fairer society.

44

CHAPTER 5

ANALYSIS OF CORUMBÁ'S URBAN AREAS ACCORDING TO THEIR LEVELS OF ENVIRONMENTAL VULNERABILITY

The process of urban growth in recent decades has potentiated and even induced some environmental problems in many Brazilian cities. According to Ross (1993), the fragility of natural environments in the face of human intervention is greater or lesser, depending on their genetic characteristics. This concept has led to many studies and discussions about the environmental problems caused by anthropic activities, especially in relation to the use and occupation of urban land and the consequent changes to the landscape, some of which are irreversible. Thus, analysing the landscape from a geosystemic perspective has been fundamental to understanding the changes caused by environmental dynamics:

> "The concept of a system is currently the best logical tool we have for studying environmental problems. It allows us to adopt a dialectical attitude between the need for analysis - which results from the progress of science and research techniques - and the opposite need for an overall vision capable of enabling effective action on the environment." (TRICART, 1977).

Currently, studies on environmental vulnerability point to an integrated way of analysing the environment, where physical and social characteristics should not be studied separately in a fragmented way. The purpose of these studies is to diagnose and classify the fragility of certain areas with regard to their use and human actions. Knowing and understanding the existence of the limits supported by nature and the restrictions imposed depending on their physical characteristics makes it possible to implement planning and definition actions for their use.

For Marandola Jr. and Hogan (2009, p.5), looking at the dangers and vulnerability of a place is a strategy that allows us, on a micro-scale, to grasp the elements that interfere in the production, acceptance and mitigation of dangers.

The identification of areas of environmental vulnerability in the city of Corumbá-MS were previously defined by Pereira and Pereira (2012) using geoprocessing techniques and based on the concept of stability of geoenvironmental units, as well as being based on the principle of ecodynamic analysis by Tricart (1977) and adapted by Costa (2006). In this way, this work uses the data provided by Pereira and Pereira (2012) to continue the studies on the environmental vulnerability of Corumbá-MS, considering the information provided by the authors and comparing it with the reality found during the field research.

The upper plateau, on the other hand, where the city centre is located, is regularly flat, with wide, tree-lined streets and good infrastructure with shops, hotels, banks, public buildings and higher-income families. At different points in the city you can find outcrops of rock, as well as the presence of swampy and pedologically unstable areas along the Paraguay River (Figure 10).

Figure 10: Rock outcrop in the city centre. **Author:** Shirley Matias, 2013

Figure 10 shows a slight slope, exposed rocks, traces of removed vegetation and trees with exposed roots, typical of shallow soils with runoff marks. The construction pattern of the house is unsafe and could be damaged during periods of heavy rain.

Based on the terrain information generated through the GIS, a thematic map was created showing the city's areas of environmental vulnerability, which associates natural vulnerability with land use occupation and serves as the basis for analysing each situation:

The water potential of the city of Corumbá is significant, because in addition to the Paraguay River, which bathes the northern part of the city, there are perennial and intermittent streams that form the urban micro-basins and make up the morphological aspect of the city. The aggravating factor in this case is the fact that their beds have been completely occupied and altered by urbanisation. Some have been channelled, plugged and even ignored for construction work. However, during rainy periods, they favour flooding, causing inconvenience to the population.

Pereira and Pereira (2010) defined 5 vulnerability classes for the city of Corumbá, which are distributed in relief units and classified as very high, high, medium, low and very low, as shown on map 02:

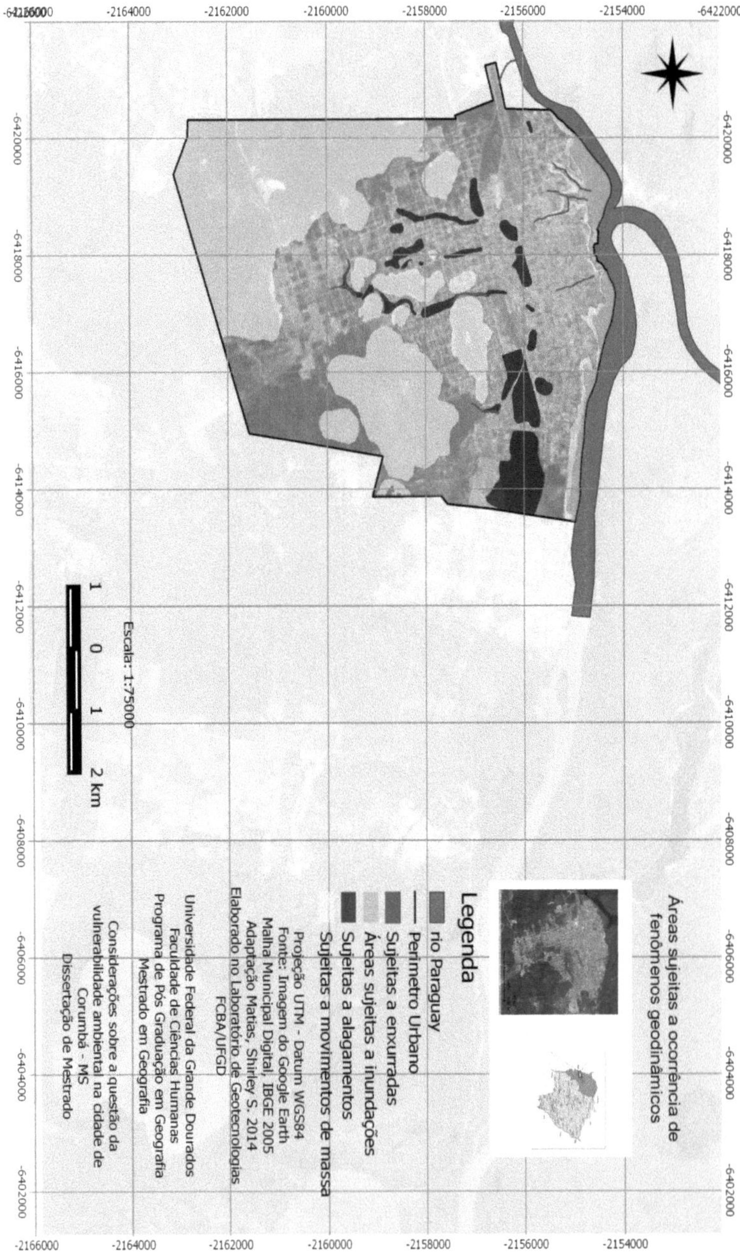

Map 02 - Areas subject to geodynamic phenomena. Source: Pereira and Pereira (2010). Adaptation: Matias;Shirley S. 2014

5.1 - Areas with low and very low vulnerability

It includes flattened areas, or areas of gentle relief, whose conditions indicate a very low probability of mass movements and do not represent a danger to use and occupation, with few restrictions on excavations or cuts. These areas are located in the upper part of the city, close to the Inselbergs, and today constitute the most stable areas of the entire urban site, with a very low population density.

The hills and hillsides in this area have dense vegetation cover and no degradation processes or erosive features resulting from rainwater runoff were found, reinforcing the importance of vegetation for soil protection and contributing to good environmental quality (figure 11).

Figure 11 - Natural vegetation of the Inselbergs in a peripheral area. Photo: Shirley Matias, 2014

5.2 - Areas with low environmental vulnerability

They indicate areas where the conditions of the soils, rocks, relief and possible existing interventions suggest a low or moderate probability of mass movements. They may have little or no evidence/records of movement. Thus, relating the physical characteristics to the vulnerability characteristics defined and presented in Table 1, these areas are located on the pediplano, an area belonging to the Tamengo Formation. This area was classified as being of medium vulnerability.

The city's commercial centre is located in this area, which is urbanised and has good infrastructure. The natural slope of the area facilitates rainwater run-off. The city centre is well wooded and the houses are built to a high to medium quality standard. Most of them are old buildings and show no signs of instability. In this area, no type of intervention was found that could potentially lead to the development of slipping or sliding processes.

Despite the stability of the area, there are still some topographical discontinuities (figure 12) that have become known by the people of Corumba as "holes" and there are a fair number of families living there.

Figure 12 - Partial view of a discontinuity in the city centre with a canal.
Photo: Shirley Matias, 2013.

Figure 12 shows the unevenness between the street and the lowered area, where a watercourse crosses, and the engineering work carried out to prevent the canal from overflowing during the rainy season and invading the surrounding homes. In a more detailed analysis, the environmental degradation found at the site could be reported on a micro scale, which could be better described and diagnosed, perhaps in a future case study.

Although it is a relatively flat and stable area, the city centre already has some drainage problems related to the sealing of the soil, which favours a greater flow of surface runoff and hinders infiltration into the soil and storage of water in the water table during rainy periods, when rainfall is more frequent and intense. *The* improper disposal of solid waste, the inefficiency of the rubbish collection and urban cleaning services provided, result in the clogging of manholes and rainwater galleries, causing inconvenience to the population during rainy periods. During the period with the highest rainfall (December to March), problems with flooding arise, the result of a population density that is out of step with the infrastructure service offered to it, especially with the precariousness of the sewage collection service that is still not in operation in the city. These are problems that are currently affecting the majority of Brazilian cities as a result of the considerable increase in the population in urban areas, which is set to increase. Planning is the determining factor so that the city is not affected by these types of problems. Investments in infrastructure and preventive measures to support growth and development in a sustainable way are essential and must be applied rigorously, thus avoiding spending public money on actions to restore heritage or support families who may be affected by flooding.

The increase in the flow of surface runoff is also reflected in other areas of the city, such as relative depressions, valley bottoms and accumulation patterns, causing soil saturation, among other problems that we'll see in the next section. These areas are prone to flooding and constitute areas of high environmental vulnerability, which will be discussed below.

5.3 - Areas with medium environmental vulnerability

Classified as areas of high environmental vulnerability, these belong to the Tamengo Formation, corresponding to part of the city centre and the banks of the Paraguay River, as well as the inselberg areas corresponding to the Bocaina Formation. However, the highlight is the areas subject to flooding.

These are areas that were occupied inappropriately during the city's growth process. The urbanisation of the city of Corumbá did not take into account the presence of the existing micro-basins on the urban perimeter, and the occupation and use of the land was superimposed on the channels and old talvegues, in order to make better use of the built-up area. The areas in question have been spatialised on the following map:

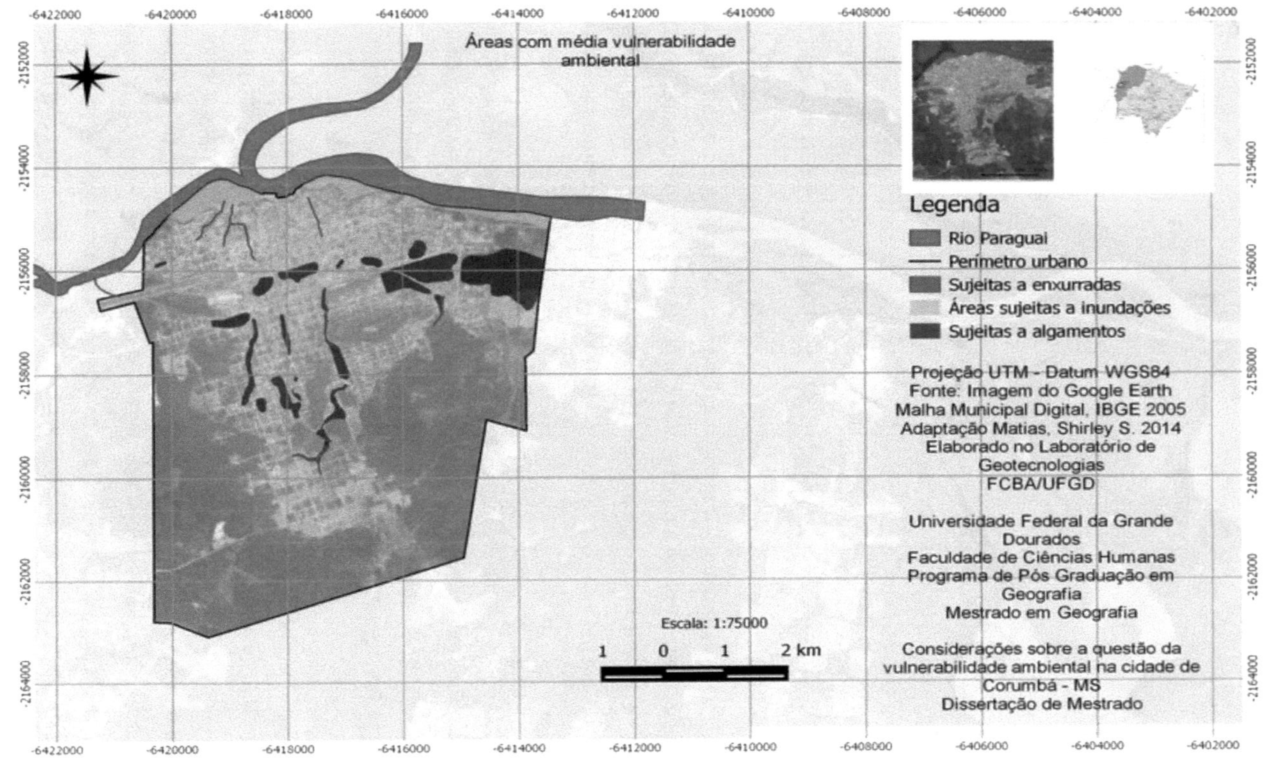

Map 03 - Areas with medium environmental vulnerability. **Source:** Pereira and Pereira (2010). **Adaptation:** Matias Shirley S. 2014.

5.3.1- Areas subject to flooding

According to Embrapa's Technical Circular No. 18, the presence of non-crystalline, porous and permeable limestone rocks favours the permanence of water in the profile throughout the wet season. The use of these areas must include technical solutions and adequate protection measures to avoid the damage caused by waterlogging and flooding, which ends up punishing residents during the rainy season. These characteristics of the limestone areas, together with the valley bottom areas, make them extremely vulnerable in the urban context of Corumbá.

The natural characteristics of these areas are further aggravated by anthropogenic actions that could not disregard the hydrographic network and urban drainage systems for conducting rainwater. Investment in infrastructure in these areas would be essential to support the demand reflected in soil sealing and the increase in housing in the area. According to Tucci (2003), the public administration has no interest in preventive action to deal with these problems because, as soon as flooding occurs, the municipality declares a public disaster and receives funds from the fund requested without the need to hold a public competition for the use of this resource. Following this logic, it is unlikely that investments will be made in structural works to solve the problem of the residents who live in these areas, who are generally from a less favoured social class, as is the case with the residents of the new housing complex built in a valley bottom area by the state government (figures 13, 14, 15, 16 and 17).

Figure 13 - Homes in a flooded area. Photo: Shirley Matias, 2014.

Figure 14 - Residence abandoned by the resident. Photo: Shirley Matias, 2014.

Figure 15 - Housing block built in a flooded area.
Photo: Shirley Matias, 2014.

Figure 16 - Housing in a flooded area. Photo: Shirley Matias, 2014.

Fig. 17- Improvised barrier to prevent water from entering; Source: Matias S. Shirley, 2014

5.3. 2- Accumulation Modelling

The Modelados de Acumulação are the flattened terrains occupied by the Pantanal formation, which extend along the entire riverside of the urban area, along the Beira Rio and Cervejaria neighbourhoods and between the Paraguay River and the riverside slopes. These areas are subject to the hydrological flood cycle of the Paraguay River and the Tamengo Canal. The soil in these areas is of the lithic type, shallow, poorly developed with a clay texture and very low porosity, limiting its infiltration potential and leading to greater surface water run-off (MONTEIRO, 1997).

Located in the lower part of the urban area, the land is affected by runoff from the higher parts of the city. The presence of buildings throughout the area increases the volume of surface water discharged directly into the riverbed, compromising the quality of its waters and polluting the water table.

The natural slope of the city, which behaves like a ramp towards this area, causes the volume and speed of water to increase, favouring the occurrence of torrents, which is intensified by the sealing of the soil, while at the same time reducing the time this water remains in the system (GUERRA, 2007).

Sediment transport and deposition in these areas also favours siltation and increases the risk of flooding. Occupation in these areas is precarious, some residents have filled in the land in an attempt to prevent water from invading during the flooding period, and the lack of basic sanitation infrastructure contributes to the degradation of water quality (Figures 18, 19 and 20) below:

Fig. 18: Housing in the Beira Rio neighbourhood. Photo: Shirley Matias, 2014.

Fig. 19: House built on the banks of the Paraguay River. Photo: Shirley Matias, 2014.

Fig. 20: Housing in the flood area. **Photo:** Shirley Matias, 2014.

Floodplains are floodplains that help control river floods by accumulating the water that overflows from watercourses until it is absorbed by the soil. These areas are unsuitable for occupation, as they are hit by floods from time to time, making their occupation an environmental problem. They are considered permanent preservation areas and deserve special attention in the city's spatial planning process.

5.3.3 - Relative Depressions

The relative depressions are located between the inselbergs of the Bocaina Formation. They are flat, well-urbanised areas and the soil is generally waterproofed by buildings and the absence of drainage infrastructure. The infiltration and drainage capacity of rainwater is minimal and favours the retention of surface runoff from the higher areas, leading to flooding during periods of significant rainfall (PEREIRA & PEREIRA, 2010).

This class is also susceptible to the falling of loose blocks on the slopes of hills, which can have serious consequences such as the loss of property and even lives in the homes located there.

5.4 - Areas with very high vulnerability

Although the municipality covers an area of approximately 65,000 km^2 , most of this area is occupied by the Pantanal, there is only approximately 39 kim2 left for the expansion of the city, which due to its physical characteristics faces serious limitations to its growth, since its geographical limits do not allow for much expansion, as it is squeezed between the Paraguay River to the north, the Morrarias to the south, the border with Bolivia to the west and to the east it is bordered by the municipality of Ladario.

The city is experiencing the problem of a lack of housing and a lack of space to build new homes. In this way, the poorest part of Corumba's population is looking for their own spaces in areas considered unsuitable or inadequate for housing, creating risky conditions for themselves in their daily lives.

The class classified as very high vulnerability includes areas that are generally unsuitable for human activities, as the terrain conditions are extremely favourable for triggering landslides, even under natural conditions. In general, they correspond to highly sloping terrain, typical of rugged regions, where even the adoption of occupation methods equipped with state-of-the-art technological resources does not eliminate the situation of imminent risk, requiring that any use of the area can live with a significant level of risk. Measures to prevent and mitigate accidents should therefore be considered.

In Corumbá, these areas considered to have the highest degree of environmental vulnerability are located on the riverside escarpments belonging to the Tamengo Formation, and on the inselbergs of the Bocaina Formation (hill slopes). These areas are spatialised on map 04, a cartographic product that shows the riverside slopes located to the north of the city and also distributed throughout the east-west extension of the city, very close to the Paraguay River and the city centre. In the south and west are the inselbergs, whose altitudes vary between 150 and 450 metres, with slopes of more than 30°, contrasting with the flat areas in the surrounding area.

The extreme vulnerability of these areas is mainly due to the geological constitution of the terrain combined with the geomorphological conditions and added aggravating factors resulting from anthropogenic actions. The steep slope of these areas makes the environment vulnerable to the severe incisions imposed by the linear surface runoff of rainwater and mass movement. These are areas that have been occupied by the poor, and the housing installed very close to the slopes exposes the inhabitants to the risks of landslides and blocks of loose rock worn away by the action corresponding to the removal of the vegetation cover from these environments, which has contributed to accentuating the fragility of poorly developed, shallow soils, with a clay texture and very low porosity, with severe limitations to their infiltration potential (MONTEIRO, 1997).

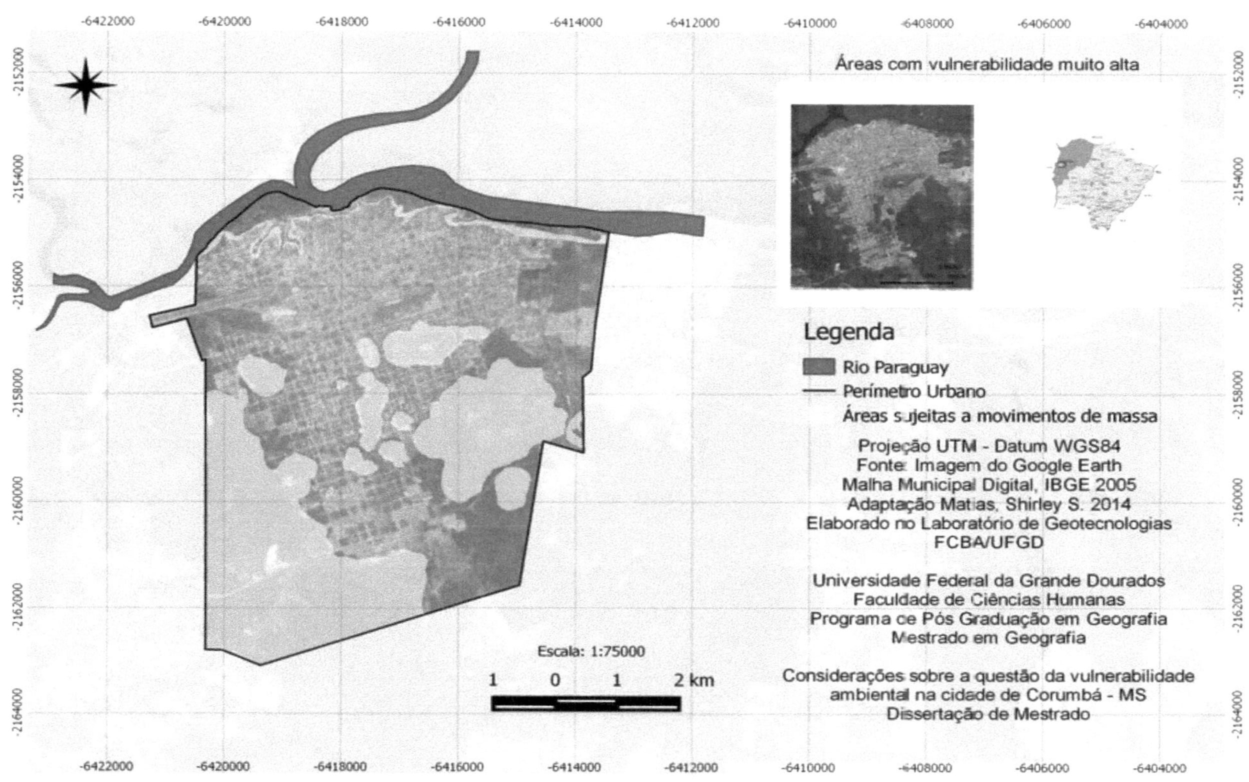

Map 04 - Spatialisation of areas with very high Environmental Vulnerability. Adaptation: Matas; Shirley S. 2014

5.4.1- Occupations on the slopes of Morro do Cruzeiro

On the slopes of Morro do Cruzeiro, occupation has occurred not only irregularly, but also in a way that has been induced and, to a certain extent, encouraged by the government itself. These are houses with a low standard of construction on land with an approximate slope of 40° and without any engineering support to prevent possible accidents, as the following images show:

Figure 21 - Housing in a hillside area. Photo: Shirley Matias, 2013.

The image in figure 21, located on the southern face of Morro do Cruzeiro, shows the situation of the terrain and the low standard of construction of the house, which has a slight slope, characterising the movement of the terrain. The exposed rocks reveal litholic soil, extremely shallow, with low infiltration capacity, which greatly favours surface run-off and the probable transport of sediment.

The colluvial sediments from the ferruginous sedimentary surfaces, which are deposited on the limestone slopes, form extremely rocky, stony and gravelly soils. The picture also shows the cultivation of banana trees, a plant that retains a large amount of water and whose roots do not structure the soil and contribute to its sliding.

Figure 22: Occupations in Morro do Cruzeiro. Photo: Shirley Matias 2013

Figure 22 shows the northern face of Morro do Cruzeiro. The image reveals the significant slope and the induced occupation of the hill by the government itself, since the power pylons are installed right up to the top of the hill. The construction of the staircase is also a way of encouraging the population to settle in the area. The presence of spreading undergrowth is the result of the removal of natural vegetation, which over the years has been replaced by medium and large trees that could fall.

The removal of natural vegetation leaves the soil exposed, facilitating the transport of the topsoil and favouring the movement of the ground, putting homes at risk.

Figure 23: Creeping vegetation on the hillside. Photo: Shirley Matias, 2013.

Figure 23 shows the undergrowth spreading as a result of the removal of natural vegetation. The image also shows the irregular cut and the proximity of the building to the hillside. The house, in turn, does not have a safe and appropriate technical construction for the type of terrain. In the event of a landslide, the house will be affected by the material from the hillside, which could cause material and human losses. Accumulated rubbish, deforestation and soil exposure are therefore factors that compromise the stability of the slope and favour surface run-off. In this way, water runs off more quickly, reaching the lower areas and carrying sediment along the way, which is when manholes become clogged and consequently flooded, depending on the volume of rainfall.

Figure 24: Hydraulic installation. Photo: Shirley Matias, 2013.

The image corresponding to figure 24 shows the extremely rocky, stony and gravelly soil, with the scars of surface runoff, revealing the low infiltration capacity. The drainage system is precarious and insufficient, with no engineering techniques to support it. No surface water drainage works were seen in the area, and the area does not have a sewage collection service, nor is it served by a rubbish collection service. However, the plots have all been regularised by the Corumbá City Hall, as mentioned in the previous chapter.

The *on-site* survey of the areas with the highest degree of vulnerability, located in the inselbergs in the southern and western sectors of the city, revealed that Morro do Cruzeiro is the one that has suffered the most from changes caused by human activity. This is probably due to its proximity to the city centre. The whole area around this hill, which is also a tourist attraction in the city and where the Cristo Rei do Pantanal was built, is occupied by housing. The improper occupation of these areas accelerates and extends the process of instability of the slopes and their natural dynamics, as these areas show the removal of the original vegetation cover, the presence of septic tanks, poor foundation construction and the dumping of rubbish on the slopes. The solution to the situation found in Morro do Cruzeiro depends mainly on the efforts and interest of the public authorities to regularise the area, either by removing the families living there or by installing infrastructure to mitigate the negative impacts.

Technical monitoring by trained professionals, the planning and implementation of works to contain the slopes, the execution of basic sanitation actions, surface drainage works and, above all, the control of occupation of these areas, are actions that must be taken urgently to mitigate the process of soil degradation and the damage caused by the water during the rainy season.

The city council's proposal to remove this population to build a park is an alternative to recovering the hillside and improving the urban landscape. However, it could take a few years and, in the meantime, if there is no supervision and control of the occupation, the tendency is for the situation to worsen, enabling the formation and expansion of favelas, as is the case in many Brazilian cities.

5.4. 2- River slopes

According to Isquierdo (2010), the escarpments of the riverside slopes were formed from the rupture, fracture and detachment of two contiguous blocks of the sedimentary package of the Corumbá Group. These slopes run along the entire riverside in an east-west direction from the urban site. These escarpments have easily erodible rocks, with the presence of exposed roots and fractures in the rocks, which are directly impacted by the action of rainwater, making it possible for cliffs to fall during periods of heavy rainfall.

The vegetation on some parts of the slopes is non-existent, favouring landslides and, in other areas, the presence of large trees with exposed roots poses an imminent risk of toppling over, with the risk of hitting the houses located on the lower level. The presence of large amounts of rubbish scattered across the slopes and in the backyards of homes is an aggravating factor, as it favours the concentration of water inside the soil, since the rocky substrate is very porous and favours soil saturation. Buildings are present along the entire length of the slopes, on both levels. The houses on the lower slopes have a poor standard of construction and no infrastructure whatsoever.

In some places, the houses on the upper level have a high standard of construction and good infrastructure, revealing the owner's high purchasing power, in contrast to the majority of the houses. The various situations encountered during the tour of this area will be detailed below, taking as an example the intersection of Ladeira d. Emília Giordano and Rua Manoel Cavassa, in the Beira Rio neighbourhood, where the most aggravating situations are found, as shown in the following sequence of photos: (25, 26 e 27)

Figure 25 - Occupations on the riverside slopes in the Beira Rio neighbourhood. Photo: Shirley Matias, 2013.

Figure 26 - Collapse of a hillside block in the Beira Rio neighbourhood. Photo: Shirley Matias, 2013.

There are houses of a low construction standard, fragile, located extremely close to the hillside where the vegetation has been removed by the residents. Figure 26, for example, shows the existence of a block of rock that has come loose from the hillside and shows that the house is at high risk, susceptible to the collapse of loose blocks due to the fragility of the terrain, which is easily erodible, with exposed roots and little or no vegetation. The rocky substrate is very porous and favours saturation of the slope.

Figure 27 - Slope with fragile, porous soil that favours infiltration. Photo: Shirley Matias, 2013.

Figure 27 shows a medium-sized tree with exposed roots and a risk of toppling. The residence on the upper level is already very close to the slope. It is possible to see the fragility of the soil, which shows cracks and fissures and favours the infiltration of water, which in this case, due to the porosity of the rock, remains inside for longer, favouring saturation and contributing to increasing its fragility. The removal of the hillside's natural vegetation is also aggravating in this case, as it leaves the soil completely exposed. An interesting fact that happens in this area is that the residents not only remove the vegetation, but they also have the habit of digging up the hillside so that it is far away from their houses. This fact was recounted by one of the residents during the field research data collection.

Figure 28 - **Housing in a risk area.** Photo: Shirley Matias, 2013.

Figure 28 shows the houses built close to the slope, which shows traces of slippage, although the uncovered area is small. The situation in this case presents a double risk, as there are houses on both levels and there is an overload on the slope, favouring instability. There is also rubbish dumping and the discharge of waste water from the houses on the upper level into the houses below as a result of the inefficiency of the rubbish collection service (figure 28).

Figure 29- Disposal of household rubbish on the hillside. Photo: Shirley Matias, 2013.

Figure 30 - High-standard building at the top of the slope. **Photo:** Shirley Matias, 2013.

In the case of figure 30, part of the limestone rock may have crumbled due to the growth of tree roots along its fractures, causing blocks to fall. The action of rainfall and the discharge of waste water from homes also favours the weakening of the slope.

The conditions recorded *on site in* the areas considered to have a very high degree of environmental vulnerability show the great susceptibility to which the families living in the area are exposed, with a great risk of loss of life. Preventive measures must be taken as a

69

matter of urgency to guarantee the safety of this population, in which case the best prevention would be to relocate these families to more appropriate and risk-free areas.

CHAPTER 6

FINAL CONSIDERATIONS

A review of the literature on the city of Corumbá reveals that certain problems have accompanied the city's history. Occupations in areas unsuitable or inappropriate for housing began shortly after the Paraguayan War and continue to this day.

The neglect of the city and the poor population are also widespread situations; social problems are intertwined with environmental problems and unfortunately these facts are ignored. Corumbá, 234 years old, has not yet had the good fortune to be honoured with an administration that really cares about its people, its ways and its sustainable development.

The drafting of the Participatory Master Plan was a great achievement for the city, although it is still not being applied in full. A review and update of municipal legislation is an action that should be taken as soon as possible. The Land Use and Occupation Law needs to be created as a matter of urgency, as it currently doesn't exist, with the municipality relying solely on a few guidelines from the Municipal Code of Postures.

Municipal environmental laws also need to be put in place, as the problems are many and a city whose greatest asset is the environment itself cannot neglect the many aggressions it has suffered. It is worth emphasising that working together with the residents of the most vulnerable areas would be of great value, as they live with the risks of an imminent tragedy.

The town hall and Civil Defence could promote debates, seminars and other ways of discussing and advising on the dangers and even safety measures while the situation is not definitively resolved.

Through this work, it was possible to see that the processes of instability in the verified areas are due both to natural characteristics, which in themselves are fragile, and to anthropic actions combined with a lack of urban planning. The field research carried out in the study area made it possible to reinforce and detail the environmental vulnerability existing in the city of Corumbá-MS, diagnosed by Pereira and Pereira (2010), using remote sensing resources. During the fieldwork, several environmental problems were found, such as: soil degradation, deforestation, occupation of hillside areas and valley bottoms, and disregard for water resources, are just some of the environmental impacts that were verified, impacts that may be irreversible.

In the course of the research, it was observed that environmental legislation, the use and occupation of municipal land and the masterplan are not being applied and, in some cases, even ignored. The inappropriate practice of setting up homes in unsuitable areas has become commonplace in several Brazilian cities and the reflection is also notorious today, making it essential to rethink the relationship between man and nature.

Preventive and mitigating measures need to be drawn up to protect and conserve the water resources in the urban area, just as the slopes need more attention and supervision to control occupation, as the data from the study shows. Likewise, the removal of families living in areas with high and very high vulnerability must be carried out urgently, as the conditions to which they are exposed indicate possible aggravations and tragedies.

With regard to the areas with low and very low vulnerability that still have good environmental quality, it is important that the expansion of the city into these areas is very well planned, applying the legislation regarding hillside areas as APP'S - Permanent Preservation Areas so that the equilibrium conditions that currently exist are preserved.

In this sense, it is necessary for public authorities to rely on a multidisciplinary team of professionals, so that the city's growth can be planned in an integrated manner that can take into account all the natural features in order to re-establish a harmonious, healthy and sustainable environment.

BIBLIOGRAPHICAL REFERENCES

AB'SABER, Aziz N. The Domains of Nature in Brazil: landscape potential. São Paulo: Ateliê Editorial, 2003. 159 p.

ARAÚJO, L. A. *Environmental Expertise in Public Civil Actions.* Bertrand Brasil, 2010.

BASTOS, S. C. A. and FREITAS, C. A. *Agents and Processes of Interference, Degradation and Environmental Damage in Avaliação e Perícia Ambiental.* Bertrand Brasil, 2010.

BARBOSA, G. A. e COSTA A. A. O *solo urbano e a apropriação da natureza na cidade.* Soc. & Nat., Uberlândia, ano 24 n. 3, 477-488, set/dez. 2012.

BITTAR, O. Y. *Technical Manual for hillside occupation* (IPT Report 28.215). 1990.

BRAZIL. Constitution (1988). *Constitution of the Federative Republic of Brazil.* Brasília, DF:

Senado Federal: Centro Gráfico, 1988. 292 p.

BRASIL, Law No. 6.938. *on the National Environmental Policy,* Brasília, DF, 31 August 1981.

BRAZIL, Law No. 10.257 of 10 July 2001. City Statute. *Regulates Articles 182 and 183 of the Federal Constitution, establishes general guidelines for urban policy and makes other provisions.*

CARLOS, Ana Fani Alessandri. *The city.* 2 ed. São Paulo: Contexto, 1994

CARLOS, Ana Fani Alessandri. *São Paulo: from industrial capital to financial capital.* In: CARLOS and OLIVEIRA (eds.). Geographies of São Paulo: the metropolis of the 20th century. São Paulo: Contexto. 2004. pp. 51-84.

CAMARGO, L. H. R. The *Geostrategy of Nature.* Bertrand Brasil, 2012.

CHRISTOFOLLETTL *Modelling Environmental Systems.* Edgar Blucher, 1999.

CORRÊA. R. L. *O Espaço Urbano.* São Paulo: Editora Ática. 1989.

CORRÊA, Valmir Batista. *Corumbá: Land of struggles and dreams.* Edições do Senado Federal - Vol.77, Brasília, 2006.

CUNHA, S. B. E GUERRA, A.J.T. *Avaliação e Perícia Ambiental.* Rio de Janeiro: Bertrand Brasil. 2003.

CURADO, Fernando Fleury. *Socio-Economic and Environmental Considerations Related to the "Arrombados" in the Taquari River Plain, MS / Fernando Fleury Curado.* - Corumbá: Embrapa Pantanal, 2004. 33p.; 16 cm. (Documentos / Embrapa Pantanal, ISSN 1517-1973; 67)

EMBRAPA. Pantanal Agricultural Research Centre - CPAP (Corumbá, MS). *Low intensity reconnaissance survey of the soils of the municipality of Corumbá and Ladário.* Technical report / Embrapa Pantanal / Rio de Janeiro: Embrapa Pantanal, 2010.

FERRARA, Lucrécia D'Alessio. The *Peripheral* Look*: Information, Language, Environmental Perception.* São Paulo: Edusp, 1993.

FOLADORI, G. *Metabolism with Nature.* Unicamp, 1999.

GODOY, Arilda S. *Introduction to qualitative research.* 1995. Available at http://www.scielo.br/pdf/rae/v35n2/a08v35n2.pdf accessed on 06 February 2014.

GUERRA, A. J. T. , ARAÚJO, G. H. S" ALMEIDA, J. R. *Gestão Ambiental de Áreas Degradadas.* Ed. Bertrand Brasil. Rio de Janeiro, 2005.

HANY. Fátmato Ezzahra Schabib. *Corumbá, Pantanal of Mato Grosso do Sul: Periphery or Central Space.* Master's Programme in Population Studies and Social Research - National

School of Statistical Sciences - ENCE. Rio de Janeiro, 2005.

HARVEY, David. Social Justice and the City. Translation: Armando C. da Silva. São Paulo: Hucitec, 1980.

ISQUIERDO, S. W. Gomes, The *Relief of the Urban Site of Corumbá.* VI Latin American Seminar on Physical Geography. University of Coimbra, 2010.

ITO, Claudemira A. *Corumbá: City Space Through the Ages.* Campo Grande-MS: UFMS, 2000.

KABIYAMA, Masato, *et al. Natural disaster prevention - basic concepts.* Curitiba: Ed. Organic Trading, 2006.

MUNICIPAL LAW 2.275 of 14 November 2012, *adds a single paragraph to art. $3°$ of Law no. 2.097.* Corumbá City Hall. DIOCORUMBÁ, 19/11/2012.

LEITE, Ricardo R. F. *Instrumentos para efetivação de políticas públicas ambientais no município de Corumbá MS.* Master's dissertation. Federal University of Mato Grosso do Sul. Aquidauana, 2008.

MARANDOLA, J; HOGAN, D. J. Natural *Hazards*: The *Geographical Study of Risks and Perils.* Campinas, 2004.

MATIAS, Shirley S. *Urbanisation and Socio-Environmental Problems: An Approach to Corumbá - MS.* XXI Encontro Sul - Mato-Grossense de Geógrafos; V Encontro Regional de Geografia. UFGD, 2013.

MONTEIRO, C. A.F. - *Geosistemas - the history of a search.* São Paulo: Contexto, 2000.

MORETTI, E. C. *Problemática ambiental no urbano: análise da ocupação do espaço e sua relação com a natureza no pantanal (Corumbá).* Master's dissertation. Faculty of Science and Technology, São Paulo State University "Júlio de Mesquita Filho". Presidente Prudente, 1996.

OLIVEIRA, Lívia de. *The perception of environmental quality.* Geography Notebooks. Belo Horizonte: PUC Minas, v. 12, n. 18, 2002, p. 29-42.

PAIXÃO, Roberto Ortiz. *Tourism on the Border, Identity and Planning a Region.* Campo Grande: Ed. UFMS, 2006.

PARTICIPATORY MASTERPLAN OF CORUMBÁ - MS. *Complementary Law No. 098/2006, of 09 October 2006.* Corumbá/MS: Corumbá City Hall, 2006.

PEREIRA, J.G. O *Património Ambiental Urbano de Corumbá MS.* Doctoral Thesis. Faculty of Philosophy, Letters and Human Sciences, University of São Paulo. São Paulo:2007.

PEREIRA, Luciana Escalante & PEREIRA, Joelson Gonçalves. *Identifying and Analysing Areas of Environmental Vulnerability in the City of Corumbá (MS).* **Geografia (Londrina),** Londrina v, 21, n.1, p, 085 - 101, jan./abr., 2012.

DIRECTOR PLAN REPORT available at http://web.observatoriodasmetropoles.net/planosdiretores/images/relatorio_corum_ba.pdf accessed on 22 January 2014.

REIGOTA, Marcos. *Ecology, elites and intelligentsia in Latin America: a study of their social representations.* São Paulo: Annablume, 1999. 115 p.

REIGOTA, Marcos. *Environment and social representation.* 7ª ed. São Paulo: Cortez, 2007.

87p.

REIGOTA, Marcos. What is environmental education? São Paulo, Brasiliense, 1991.

RODRIGUES, Aríete M. *Moradia nas Cidades Brasileiras.* São Paulo: Contexto, 1994.

RODRIGUES, C. *Applied geomorphology. Evaluation of environmental and physical-territorial planning instruments in Brazilian experiences.* São Paulo, 2v. PhD thesis - Faculty of Philosophy, Letters and Human Sciences, University of São Paulo, 2001.

SANTOS, M. *Técnica espaço e tempo.* São Paulo: Hucitec, 1994.

SOUZA, João Carlos de. *Tensions of Modernity in Corumbá.* ANPUN - XXII National History Symposium - João Pessoa, 2003.

SPOSITO, Maria E. Beltrão. *Capitalism and Urbanisation.* São Paulo: Contexto, 1994.

TRICART, J. -Ecodynamics. Rio de Janeiro: FIBGE, Planning Secretariat of the Presidency of the Republic, 1977.

TUCCI, C.E.M., *Flooding and Urban Drainage, In: Urban Flooding in South America.* Porto Alegre: Brazilian Water Resources Association, 2003. p. 45-150.

VEYRET, I. *Os Riscos; O Homem Como Aggressor e Vítima do Meio Ambiente.* São Paulo: Context, 2007.

ANNEXES

INTERVIEW SCRIPT

1. The problem of low-income families in sectors with high environmental vulnerability in Corumbá is already a historical problem. What are the city council's plans for resolving this issue?

2. With the implementation of the works related to the Growth Acceleration Programme in the city of Corumbá, what will improve in terms of infrastructure in the city area over the next twenty years?

3. The slopes of the hills (Cruzeiro, Bandeira and São Felipe), according to the Masterplan, are areas of environmental interest that are occupied by housing. Is there a project for these areas, such as the Masterplan for the Harbourfront?

4. It is well known that the city of Corumbá has problems with its urban sprawl. Thinking in the long term, what are the government's future plans for incorporating new areas of urban expansion into Corumbá?

5. What are the future prospects for the verticalisation process in the city? Are there any future plans for a possible verticalisation process?

Printed by Books on Demand GmbH, Norderstedt / Germany